Dangerous Bugs

KILLER BEES

MEGAN COOLEY PETERSON

BLACK RABBIT BOOKS

Bolt is published by Black Rabbit Books
P.O. Box 227, Mankato, Minnesota, 56002
www.blackrabbitbooks.com
Copyright © 2024 Black Rabbit Books

Alissa Thielges, editor; Michael Sellner, designer
and photo researcher

Library of Congress Cataloging-in-Publication Data
Names: Peterson, Megan Cooley, author.
Title: Killer bees / by Megan Cooley Peterson.
Description: Mankato, Minnesota: Black Rabbit Books, [2024] |
Series: Bolt: Dangerous bugs | Includes bibliographical references and index. |
Audience: Ages 8–12 | Audience: Grades 4–6 |
Summary: "Able to kill a human, killer bees are more than just a pest. Get up
close to these dangerous bugs through gross photos, leveled text, and engaging
infographics that'll make readers squirm"—Provided by publisher.
Identifiers: LCCN 2022024127 (print) | LCCN 2022024128 (ebook) |
ISBN 9781623105785 (library binding) | ISBN 9781623105846 (ebook)
Subjects: LCSH: Africanized honeybee—Juvenile literature.
Classification: LCC QL568.A6 P48 2024 (print) | LCC QL568.A6 (ebook) |
DDC 595.79/9—dc23/eng/20220630
LC record available at https://lccn.loc.gov/2022024127
LC ebook record available at https://lccn.loc.gov/2022024128

Printed in China

CONTENTS

CHAPTER 1
Killer Bee Takeover. . . . 4

CHAPTER 2
Size and Features. 8

CHAPTER 3
Where They Live and
What They Eat. 14

CHAPTER 4
Life Cycle. 20

CHAPTER 5
Watch Out!. 26

Other Resources. 30

Killer Bee

TAKEOVER

A honeybee hive buzzes. Inside, the bees feed their young and make honey. Outside, killer bees approach. They need a need place to live. And the honeybee hive will make a great spot. The queen killer bee and a few of its • • • • • workers crawl inside. The takeover has begun.

Attack!

The killer bees find the honeybee queen. They kill it. The killer bee queen then takes over the hive. It begins laying eggs. Meanwhile, the workers fight off attacking honeybees. The nest will soon be filled with killer bees.

Killer bees don't just kill honeybees. They can kill other animals, even people!

CHAPTER 2

Size and FEATURES

Killer bees look a lot like honeybees. They are about the same size. Like honeybees, killer bees have three main body parts and six legs. Two sets of wings grow from the **thorax**. Hairs on the bee's back legs pick up and carry **pollen**. A mouthpart called the **proboscis** sucks up nectar.

How Big Is a Killer Bee?

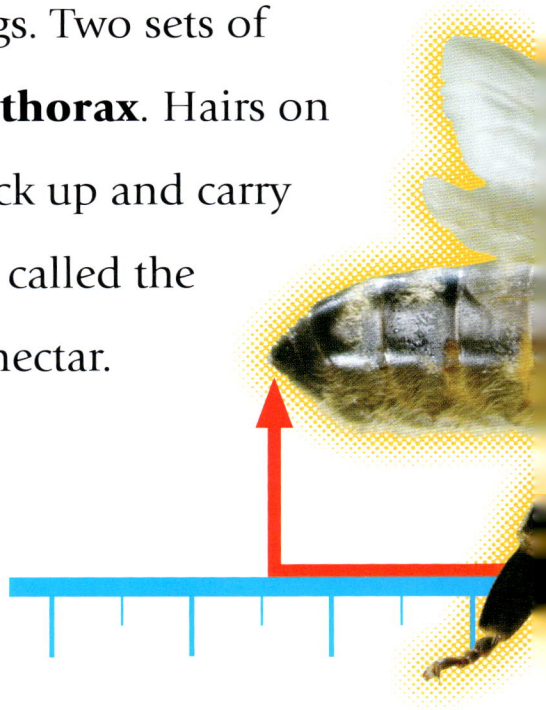

LENGTH
about **0.75**
INCH
(2 centimeters)

Killer bees do not have more venom than other bees. Larger groups of killer bees attack at once, though. They stalk threats farther too.

Serious Stingers

These fierce bees earned their name. They will stop at nothing to protect their hives from other animals. They sting with **venom**. Only females have stingers. The stingers grow from the same body parts that lay eggs. Stingers have small hooks on them. The hooks get stuck in animals and people. After the bees sting, the stingers are ripped from their bodies. The bees die soon after.

PARTS OF A KILLER BEE

HEAD

THORAX

ANTENNAE

LEGS

WINGS

STINGER

ABDOMEN

13

CHAPTER 3

Where They Live and WHAT THEY EAT

Killer bees first appeared in Brazil in the 1950s. Scientists there were working with African and European honeybees. They wanted to create bees that could make more honey and handle hotter temperatures. The scientists accidentally created a new **aggressive** type of bee. Then some of those bees escaped. They began to spread and grow.

Killer bees are also called Africanized honeybees.

Tree
Hive

Building
Hive

Where Killer Bees Live

North
America

Europe

Asia

Africa

South
America

Australia

Flying Around

Today, killer bees live all across South and Central America. They also call the southern United States home.

Killer bee colonies live in hives. They often build their hives in trees. They also take over hives from honeybees. Killer bees even build nests in old tires, cement blocks, and buildings.

The Hive

Inside their hives, killer bees build hundreds of six-sided cells. These cells, called the honeycomb, are made of wax. Each cell is like a tiny room. The young live in some cells. Other cells are used to store honey. Workers collect nectar from flowers to make honey. Colonies feed on the honey.

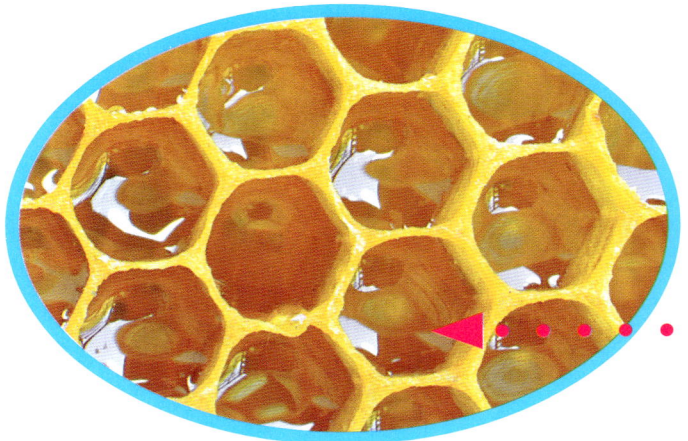

The HONEYCOMB

cells

young

honey

CHAPTER 4

LIFE CYCLE

Each killer bee colony has a queen. The queen lays all of the eggs for the colony. The young hatch as **larvae**. Worker bees care for the larvae and the queen. Killer bees grow from egg to adult in about 21 days.

Many Members

The average killer bee colony has 50,000 bees. Sometimes the hive becomes crowded. When there's not enough space, the old queen and some workers leave. They find a place to begin a new colony. This action is called swarming. A new queen takes over the hive they left.

Killer bee colonies can swarm more than 10 times a year. Honeybees only swarm two or three times a year.

From Egg to ADULT

Queens can lay up to 1,500 eggs a day. A queen lays one egg in each cell.

EGG

Adults that become queens do nothing but lay eggs. Workers care for the young and find food.

LARVA

Workers feed the larva.

PUPA

The pupa looks like a pale adult bee.

ADULT

CHAPTER 5

WATCH
Out!

Killer bees might look like regular honeybees, but they don't act like them. Killer bees get angry when people get too close. Thousands of killer bees pour out of the nest. They chase and swarm people. They sting a person's eyes, mouth, and nose. Too much venom can be deadly. It causes **organ** failure, which can lead to death. These dangerous bugs give regular honeybees a bad name.

BY THE NUMBERS

FLYING SPEED

12 to 15
MILES
(19 to 24 km)
PER HOUR

ABOUT
1,000
NUMBER OF KILLER BEE STINGS NEEDED TO KILL AN ADULT HUMAN

28

50 DAYS

average life span of worker killer bees

0.25 MILE
(0.4 kilometer)

distance killer bees will chase people

1990

year Africanized bees came to Texas

29

GLOSSARY

aggressive (uh-GRES-iv)—showing a readiness to fight, argue, or attack

larva (LAR-vuh)—the wormlike form of an animal that hatches from an egg

organ (OHR-guhn)—a structure inside the body made of cells and tissues that performs a specific function

pollen (PAHL-en)—powdery, yellow grains on flowering plants

proboscis (prah-BAH-sis)—a long, thin tube that is part of a bee's mouth

thorax (THAWR-aks)—the middle section of an insect's body

venom (VEH-num)—a poison made by animals used to kill or injure

BOOKS

Culliford, Amy. *Killer Bees.* Deadliest Animals. New York: Crabtree Publishing, 2022.

Goldish, Meish. *Killer Bees.* No Backbone! The World of Invertebrates. New York: Bearport Publishing, 2020.

Levy, Janey. *Lethal Insects.* Mother Nature Is Trying to Kill Me! New York: Gareth Stevens Publishing, 2020.

WEBSITES

Facts about Bees
www.dkfindout.com/uk/animals-and-nature/insects/bees-and-wasps/

Killer Bees
https://www.pestworldforkids.org/pest-guide/bees/#Killer-Bees

Killer Bees! | National Geographic
www.youtube.com/watch?v=d-7kKqgPEGs

INDEX

E

eggs 7, 11, 20, 24

F

features 8, 11, 12–13

H

honeybees 4, 7, 8, 14,
        15, 17, 23, 26

honeycombs 18–19

L

life cycle 20, 24–25, 29

N

nectar 8, 18

nests 7, 17, 26

O

origin 14, 29

R

range 16–17

S

size 8–9

speed 28

stinging 11, 26, 28

swarming 23, 26

V

venom 10, 11, 26